SIGHT WORDS

With Marsee

THIS BOOK BELONGS TO:

....

Tgsketch Illustration

SIGHT WORDS

Email us at
tgosketch@gmail.com
ISBN 978-1-955126-32-8

TgosketchPress
Chicago, Illinois
www.tgosketch.com

Hi, friend! It's me, Marsee!
Did you know some words are a little extra special?

They're called sight words, and we don't sound them out like other words. Nope! We learn to see them and say them right away.

Words like the, you, can, and me pop up all the time when we read. So instead of stopping to figure them out, we learn to know them by heart just like we remember a good friend's name! Some sight words are a little tricky because they don't always sound the way they look. But don't worry, with a little practice, you'll be spotting them like a pro!

Just remember:
Look at the word.
Say it right away.

Don't try to sound it out just know it by sight!

We'll learn them together, one by one.
You're getting stronger, braver, and more amazing!

Let's go find some words, okay?
Love, Marsee

Circle and Say The Word:

do
has
the
up
like
she
and
an
no
am
so
is

we
he
it
me
to
can
my
in
go
a
on
you

the

Trace the Word:

the the the the

Practice Writing:

Say the word "the" then color the word.

the

Circle and Say The Word:

so
an
can
she
to
up
has
is
on
do
he
like

me
and
you
am
the
in
no
a
it
go
my
we

we

Trace the Word:

we we we we we

Practice Writing:

Say the word "we" then color the word.

we

Circle and Say The Word:

like
the
an
on
and
go
no
a
my
to
in
so

he
she
am
it
me
we
up
do
you
is
has
can

you

Trace the Word:

you you you

Practice Writing:

Say the word "you" then color the word.

you

Circle and Say The Word:

has	on
do	in
my	and
up	to
the	like
is	she
you	me
a	can
we	it
go	no
so	an
am	he

a

Trace the Word:

a a a a a a

Practice Writing:

Say the word "a" then color the word.

a

Circle and Say The Word:

we
no
like
an
it
up
on
in
can
she
do
so

has
he
my
you
and
a
go
am
me
is
to
the

he

Trace the Word:

he he he he

Practice Writing:

Say the word "he" then color the word.

he

Circle and Say The Word:

no
to
an
the
she
we
in
it
you
am
and
he

on
like
do
so
has
go
me
can
my
up
is
a

and

Trace the Word:

and and and and

Practice Writing:

Say the word "and" then color the word.

and

Circle and Say The Word:

and
go
an
my
do
am
has
so
is
me
no
you

we
on
she
in
it
a
up
can
to
the
like
he

so

Trace the Word:

so so so so

Practice Writing:

Say the word "so" then color the word.

so

Circle and Say The Word:

am
my
it
me
a
an
on
he
has
up
and
to

we
so
go
like
is
in
can
the
do
you
she
no

up

Trace the Word:

up up up up

Practice Writing:

Say the word "up" then color the word.

up

Circle and Say The Word:

he
to
can
on
like
up
and
so
an
no
do
you

am
we
in
it
a
the
is
she
has
my
me
go

like

Trace the Word:

like like like like

Practice Writing:

Say the word "like" then color the word.

like

Circle and Say The Word:

do
is
my
like
in
she
can
to
the
am
it
you

no
so
has
an
he
on
a
we
go
me
and
up

no

Trace the Word:

no no no no

Practice Writing:

Say the word "no" then color the word.

no

Circle and Say The Word:

my	am
the	do
no	go
me	like
up	in
can	you
on	he
a	we
so	she
and	an
it	is
has	to

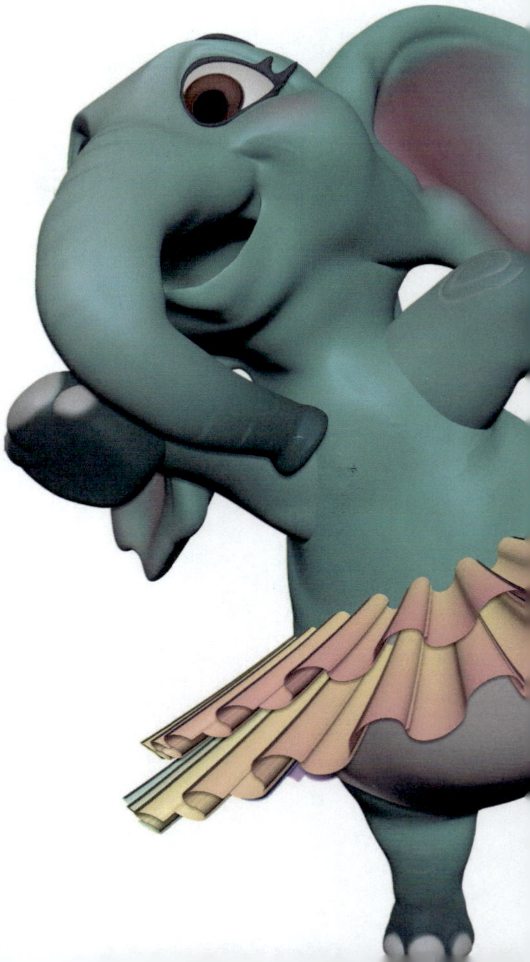

am

Trace the Word:

am am am am

Practice Writing:

Say the word "am" then color the word.

am

Circle and Say The Word:

we
the
you
no
he
on
it
and
is
an
to
so

do
like
in
has
me
am
up
can
a
my
go
she

is

Trace the Word:

is is is is is

Practice Writing:

Say the word "is" then color the word.

is

Circle and Say The Word:

the
to
has
he
a
it
do
go
like
in
am
you

and
on
can
an
we
no
me
is
so
she
my
up

on

Trace the Word:

on on on on

Practice Writing:

Say the word "on" then color the word.

on

Circle and Say The Word:

and	up
me	has
he	to
so	we
can	do
in	she
the	on
my	is
am	you
like	a
go	it
no	an

go

Trace the Word:

go go go go

Practice Writing:

Say the word "go" then color the word.

go

Circle and Say The Word:

the
it
so
he
can
up
me
go
to
no
am
we

do
has
you
an
and
a
in
like
my
she
on
is

can

Trace the Word:

can can can can

Practice Writing:

Say the word "can" then color the word.

can

Circle and Say The Word:

can	has
it	we
and	is
up	am
do	me
the	she
you	so
he	in
no	to
like	my
on	an
go	a

me

Trace the Word:

me me me me

Practice Writing:

Say the word "me" then color the word.

me

Circle and Say The Word:

it	has
she	am
in	on
we	do
me	like
a	you
is	the
to	and
no	my
go	can
so	an
up	he

it

Trace the Word:

it it it it it it

Practice Writing:

Say the word "it" then color the word.

Circle and Say The Word:

like
a
she
it
and
do
on
we
an
am
my
can

no
me
has
he
the
you
up
so
in
is
to
go

an

Trace the Word:

an an an an

Practice Writing:

Say the word "an" then color the word.

an

Circle and Say The Word:

in	can
he	has
like	up
an	she
me	and
no	go
a	is
it	the
you	do
we	am
so	my
to	on

in

Trace the Word:

in in in in in

Practice Writing:

Say the word "in" then color the word.

in

Circle and Say The Word:

up	and
a	is
to	me
in	you
it	she
on	am
no	so
we	he
like	go
an	can
has	do
the	my

has

Trace the Word:

has has has has

Practice Writing:

Say the word "has" then color the word.

has

Circle and Say The Word:

in
me
and
she
it
he
to
a
on
you
no
is

an
go
do
up
the
my
can
like
has
we
so
am

do

Trace the Word:

do do do do

Practice Writing:

Say the word "do" then color the word.

do

Circle and Say The Word:

we	am
go	on
has	so
it	is
up	and
can	the
an	me
like	you
no	he
do	my
she	in
a	to

to

Trace the Word:

to to to to to

Practice Writing:

Say the word "to" then color the word.

to

Circle and Say The Word:

no	go
it	up
you	is
can	she
has	a
in	my
do	like
am	the
me	we
on	so
he	an
to	and

my

Trace the Word:

my my my my

Practice Writing:

Say the word "my" then color the word.

my

Circle and Say The Word:

me	a
to	is
no	has
do	in
go	and
she	it
the	am
up	he
so	can
like	we
you	on
my	an

she

Trace the Word:

she she she she

Practice Writing:

Say the word "she" then color the word.

she

Made in the USA
Monee, IL
05 November 2026

67948877R00031